Awesome, Disgusting Science

Gross Science of Blood

Stephanie Bearce

Black Rabbit Books

Hi Jinx is published by Black Rabbit Books
P.O. Box 227, Mankato, Minnesota, 56002.
www.blackrabbitbooks.com
Copyright © 2026 Black Rabbit Books

Alissa Thielges, editor; Jason Knudson, designer and photo researcher

All rights reserved. No part of this book may be reproduced in any form without written permission from the publisher.

Library of Congress Cataloging-in-Publication Data
Names: Bearce, Stephanie author
Title: Gross science of blood / by Stephanie Bearce.
Description: Mankato, MN: Black Rabbit Books, [2026] | Series: Awesome, disgusting science | Includes bibliographical references and index. | Audience: Ages 8-12 | Audience: Grades 4-6
Identifiers: LCCN 2025017491 (print) | LCCN 2025017492 (ebook) | ISBN 9781645824909 library binding | ISBN 9781645824961 paperback | ISBN 9781645825029 ebook
Subjects: LCSH: Blood—Juvenile literature
Classification: LCC QP91 .B148 2026 (print) | LCC QP91 (ebook) | DDC 612.1/1—dc23/eng/20250725
LC record available at https://lccn.loc.gov/2025017491

Printed in the United States of America.

Image Credits

Alamy Stock Photo/Daniel Heuclin, 16; Freepik/7, 23, AI image creator, cover, 1, brgfx, cover, 1, EyeEm, 16, freepik, cover, 1, 10, Ksu_Ksu, 12, prettyvectors, 15, rashevskymedia, 14-15, storyset, 11, 17, 19, user1905337, 19; Getty Images/Westend61, 7; Shutterstock/ Balinda, 5, BNP Design Studio, 9, Iaremenko Sergii, 2-3, 12, Memo Angeles, 20, Miriam Gil Albert, 11, Parilov, 7, Phonlamai Photo, 12-13, SERGEI PRIMAKOV, 8, stockstation, 18-19, Stone36, 18-19, Suan Taang, 4, Svetlana_Smirnova, 7, 21, xpixel, 9, 21

Every effort has been made to contact copyright holders for material reproduced in this book. Any omissions will be rectified in subsequent printings if notice is given to the publisher.

CONTENTS

CHAPTER 1
Got Blood?............ 5

CHAPTER 2
The Experiments...... 6

CHAPTER 3
Get in on the Hi Jinx... 20

Other Resources............22

Chapter 1
GOT BLOOD?

Blood. Red and sticky. It oozes from scrapes and cuts. Some people faint when they see it. Leeches and ticks drink it. And your life depends on it. Blood carries oxygen and **nutrients** to your body. Without it, you die.

Because blood is so important, scientists are always experimenting. They want to learn more. But studying blood can be, well, disgusting.

Chapter 2
THE EXPERIMENTS

Udderly Weird Blood

Would you like cow's milk in your veins? You might try it to save your life!

People donate blood. But it is only good for 42 days. Scientists want to make **synthetic** blood that lasts longer. Doctors tried cow's milk. They **injected** it into people's veins. And you know what? It sort of worked. But it was not as good as human blood.

Adults have about 1.3 gallons (5 liters) of blood in their bodies.

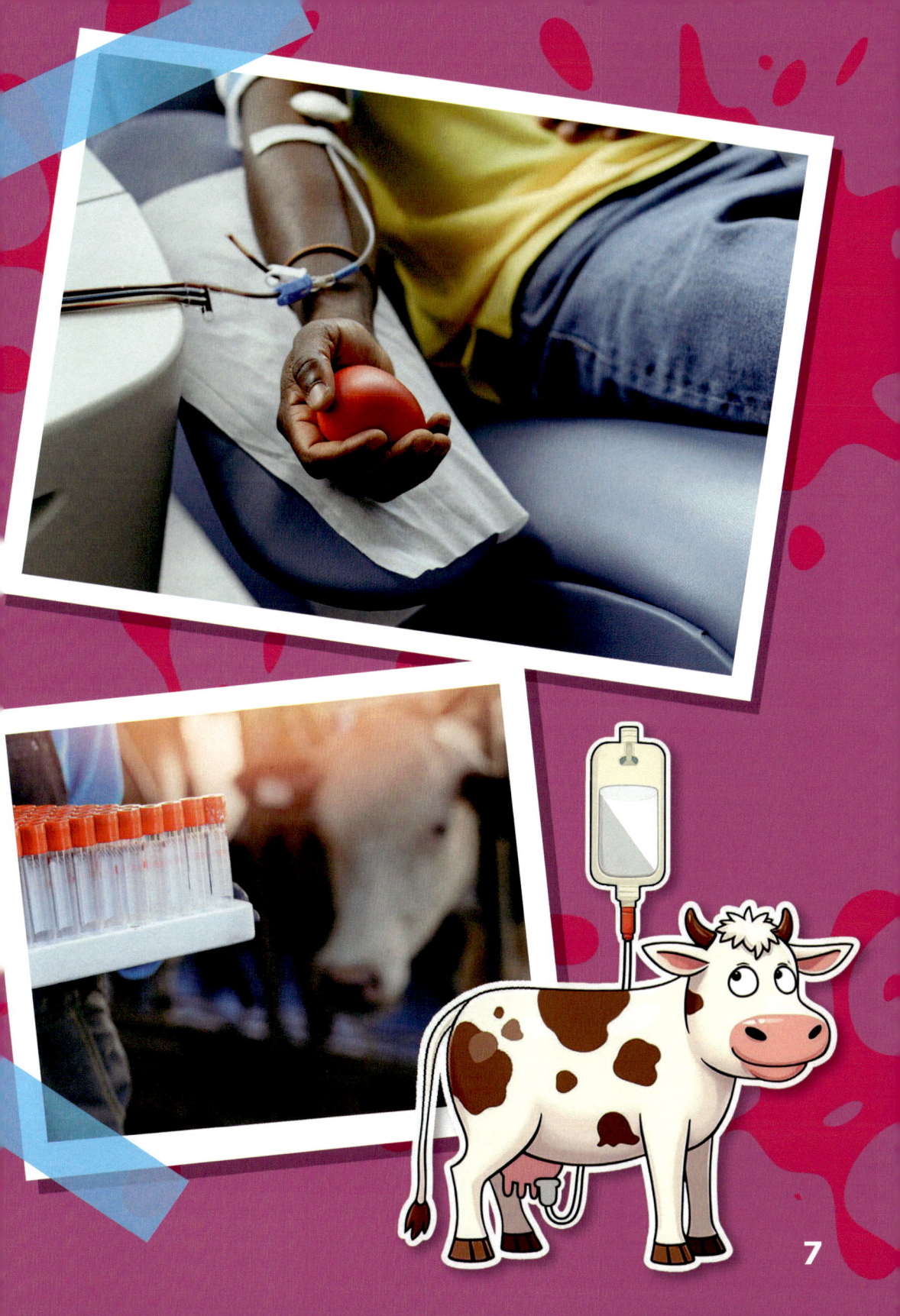

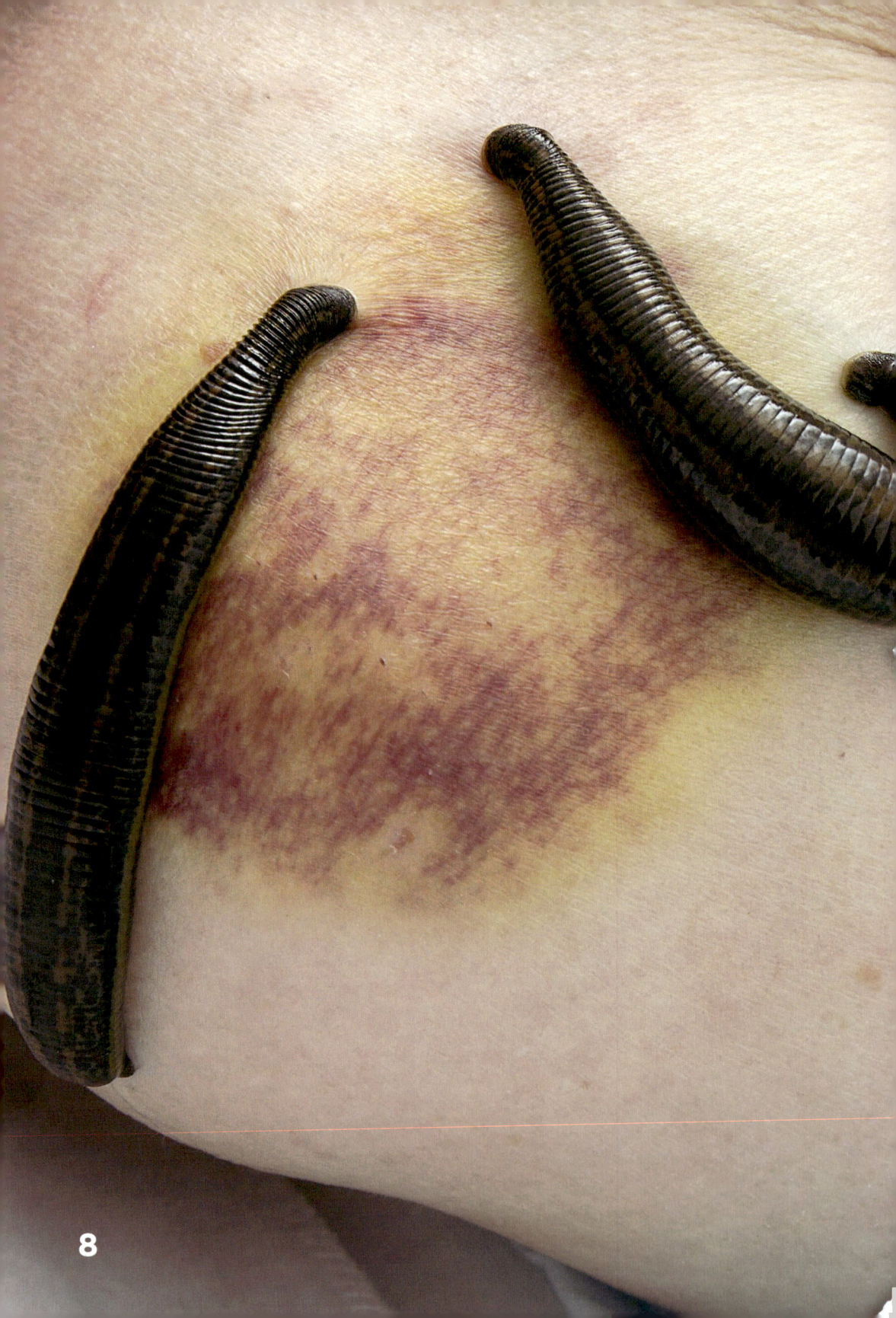

Suck It Up

The thought of leeches can make people cringe. Yet these blood-suckers are helpful for healing. Their saliva is special. It stops blood **clots** from forming. The leeches keep blood flowing to damaged areas. This helps new veins grow. The body part heals faster.

The fear of blood is called hemophobia.

Blood for Dinner

Drink your own blood? Gross! But people did it in an experiment. Were they vampires? Nope. It was for science. Scientists checked the people's poop. They looked for a **protein** found in white blood **cells**. Did they find it? Yep! The test patients had higher levels of it. The findings might help treat bleeding in the gut.

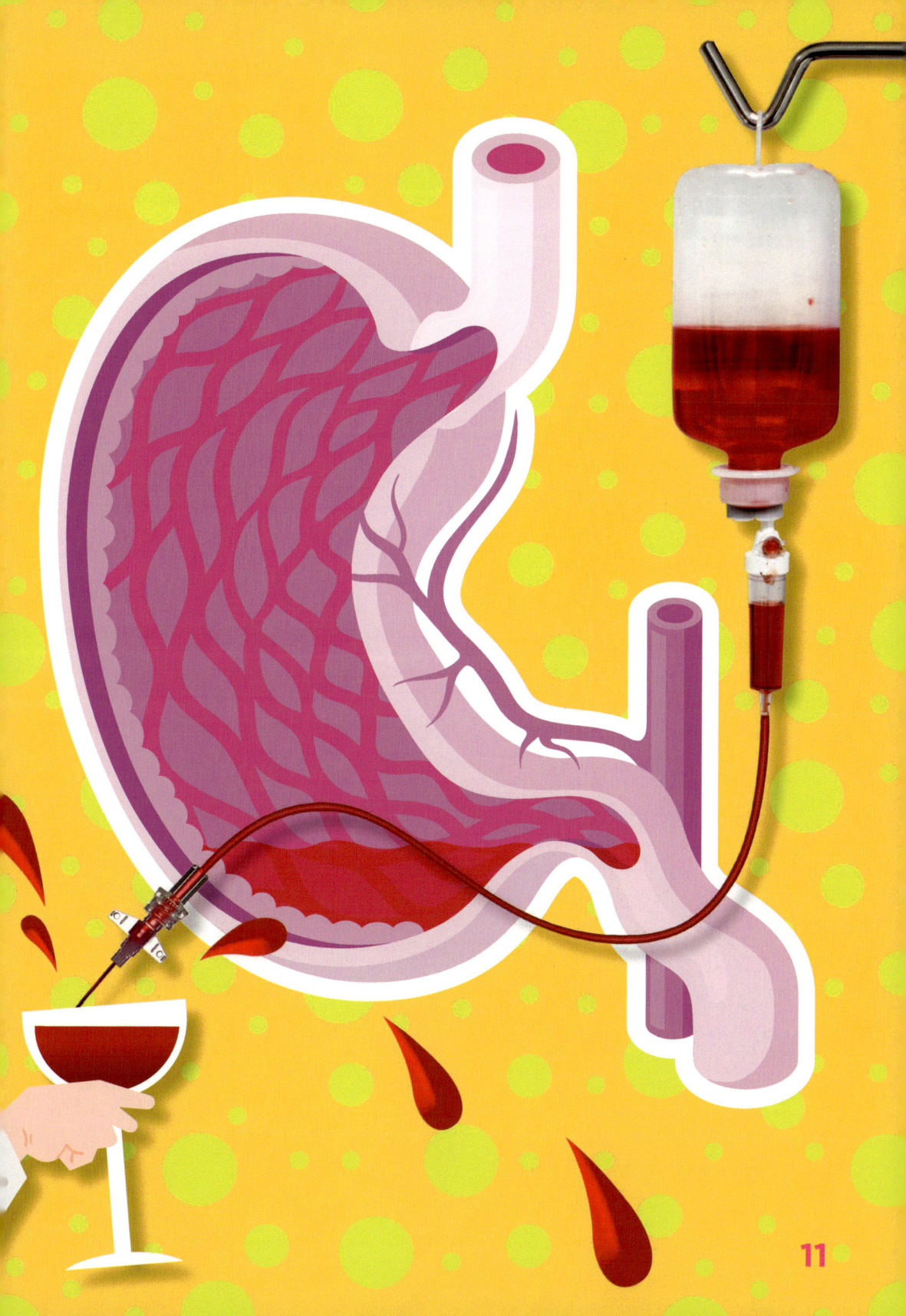

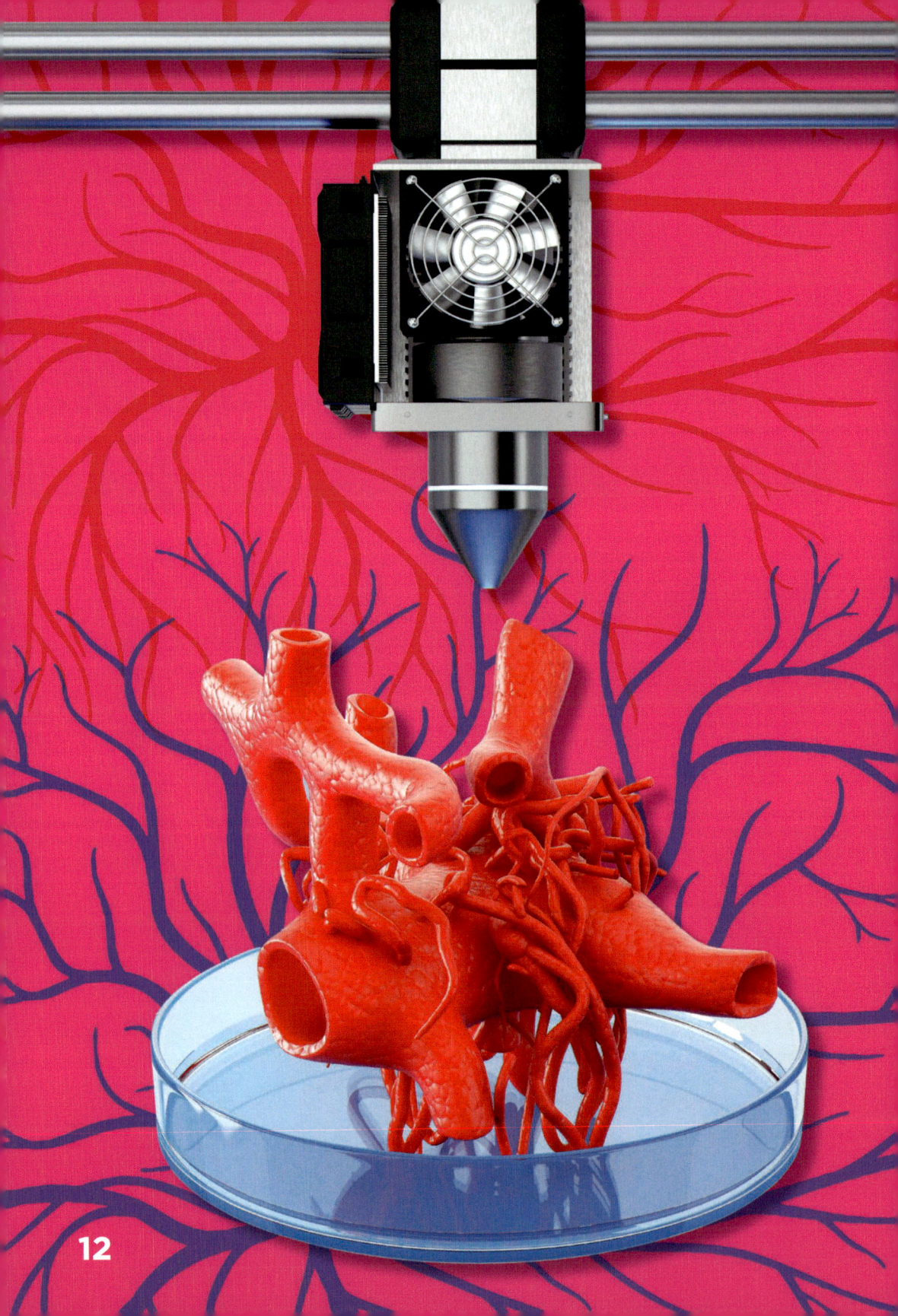

Blood in the Printer

Need new blood vessels? Doctors may be able to print them! Scientists are experimenting with this. They use living cells in a 3D printer. They make hollow tubes of cells. These can be used as blood vessels. Someday, they hope to print organs like hearts and kidneys. These can be used for **transplants**.

The human body makes about 2 million red blood cells each second.

Bear Blood Breakthrough

Could bear blood save lives? It might. Humans get blood clots from sitting too long. Not bears! Their blood does not clot during **hibernation**. Why? Their body reduces a protein that causes clots. This can be used to treat humans. People often die from blood clots. A new treatment could save lives.

BLOOD CLOT

Blood and Venom

Bleeding stopped by snake **venom**? Yesssss! Scientists experimented with proteins from two deadly Australian snakes. When combined, they create a gel. The gel forms a clot that stops bleeding. It can save lives in war zones.

Scorpions have blue blood. Some lizards have green blood. And an octopus has purple blood!

Frozen Blood

Can blood come back to life? Scientists froze human blood for years. When they thawed it, some of the blood cells started moving again. It was like they were waking up from a long sleep. How cool! Frozen blood could be used for emergencies or space travel. Someday, it might save lives!

Chapter 3
GET IN ON THE HI JINX

Imagine printing a new heart or pair of kidneys. You could save lives with high tech. To do this, become a 3D printing technician. You'll need to study math and science. After high school, you will go to a technical school or college. Then get a job in a science lab. Who knows? One day you may even print blood.